# AI and DECEPTION

## Plagiarism, Deepfakes, and More

Carla Mooney

San Diego, CA

Printed in the United States

**For more information, contact:**
ReferencePoint Press, Inc.
PO Box 27779
San Diego, CA 92198
www.ReferencePointPress.com



LIBRARY OF CONGRESS CATALOGING-IN-PUBLICATION DATA

Author: Carla Mooney
Title: AI and Deception: Plagiarism, Deepfakes, and More
Description: San Diego, CA : ReferencePoint Press, 2026.
Includes bibliographical references and index
Identifiers: LCCN 2025014585 (print) | ISBN
9781678210663 library binding | ISBN 9781678210670 ebook

For compete cataloging-in-publication data please go to www.loc.gov.

# CONTENTS

# A Realistic Deception

One day in 2024, Anthony's phone rang. When he answered, the voice on the line sounded exactly like his son. The young man was upset and told Anthony that he had been in a car accident and was in trouble. He explained that another person had been hurt in the accident, a pregnant woman who was taken to the hospital. Anthony and his son talked briefly before ending the call. Not long after the first call, Anthony's phone rang again. This time, a person named Michael Roberts, claiming to be a lawyer, informed Anthony that his son had been taken to jail and asked Anthony for money to pay for bail. "He said, 'You need to get $9,200 as fast as you can if you want your son out of jail. Otherwise, he's in for 45 days,'"[1] recalls Anthony.

Anthony attempted to reach his son again, but his call went straight to voicemail. He assumed that meant his son had been arrested. Anthony went to the bank and withdrew the bail money. Once he had the money, Anthony had his daughter call Roberts back. Roberts told Anthony that he would send an Uber driver and instructed the worried dad to hand over the bail money to the driver. When the car arrived, Anthony's daughter handed the bail money in a manilla envelope to the Uber driver.

After the Uber driver left, Anthony's phone rang again. This time the caller identified himself as Mark Cohen, another lawyer involved in his son's case. Cohen informed Anthony that the pregnant victim had died in the hospital. Then he told Anthony the woman's death meant his son's bail had increased to $25,000. Desperate to help his son, Anthony returned to the bank, withdrew more money, and handed it to a second Uber driver.

Meanwhile, Anthony's daughter searched for information about the car accident online. When she could not find anything, she became suspicious. She told her father she believed he had been scammed. Anthony was shocked because the first caller had sounded exactly like his son, but he realized his daughter was right. Anthony called the police.

## Old Scams, New Tech

Scams like the one Anthony experienced are not new. For many years, criminals have been attempting to trick vulnerable people by telling them a loved one needs money because they are injured or in jail. In recent years, however, scammers are using new technology to make their scams appear more realistic. "The scammers are just becoming more clever and sophisticated," says Los Angeles Police Department detective Chelsea Saeger. "They are using social media and technology to craft these very believable and convincing stories, and people really do believe they're talking to a grandchild or a government official."[2]

Artificial intelligence (AI) has made it easier for scammers to clone a person's voice and create a realistic-sounding conversation. Scammers only need a few seconds of a person's voice to feed it into an AI cloning tool. Saeger explains how easy it is for scammers to get a recording of a person's voice. "They call, and when you answer, and it's a scammer, there's silence," Saeger says. "They want you to say 'hello' or 'is anybody there?' All they need is three seconds of your voice to input it into AI and to clone it."[3] Beyond a live call, scammers can also use social media accounts to get information about someone they can use in a scam. "They'll go through your video posts, and if you or a loved one are speaking, they can grab your voice that way,"[4] Saeger says. Once scammers have audio of a real person's voice, they

> **"Scammers can pull pieces of a person's real voice and have an AI tool use those voice patterns to create synthetic conversations, copying and manipulating your voice."[5]**
>
> —Sean Murphy, chief information security officer at Boeing Employees' Credit Union

input it into AI voice cloning tools and then type words for the clone to say.

In 2023 Americans lost an estimated $2.7 billion to imposter scams, according to the Federal Trade Commission. Imposter scams include scammers pretending to be a loved one in trouble, the government, a bank's fraud department, or even a technical support expert. "Scammers can pull pieces of a person's real voice and have an AI tool use those voice patterns to create synthetic conversations, copying and manipulating your voice,"[5] says Sean Murphy, chief information security officer at Boeing Employees' Credit Union.

Security experts recommend being cautious when a loved one calls with an urgent financial need. They recommend calling the person back using a known phone number or contacting other family or friends to verify the situation. Experts also recommend setting up a family safe word or phrase that is not easy to

*AI has made it easier for scammers to clone a person's voice. Criminals trick vulnerable people into thinking they are talking with a loved one who is in trouble and needs money.*

guess and will help establish who is on the other end of a phone call. In addition, being skeptical of unsolicited phone calls, emails, and texts and refusing to share personal information with unverified individuals and businesses can help protect a person from AI imposter scams.

## A Powerful Tool for Deception

AI is a transformative technology with many positive applications, from improving customer service to helping in medical diagnosis. Yet in the wrong hands, AI has also become a powerful tool for unscrupulous individuals intent on deceiving and scamming others. Voice scams, deepfake photos and videos, plagiarism, phishing scams, and identity theft are some ways AI can be used unethically and criminally.

> **"As technology continues to evolve, so do cybercriminals' tactics. Attackers are leveraging AI to craft highly convincing voice or video messages and emails to enable fraud schemes against individuals and businesses alike."[6]**
>
> —Robert Tripp, FBI special agent in charge

In May 2024 the Federal Bureau of Investigation (FBI) warned individuals and businesses nationwide about the increasing threat from criminals using AI. The FBI noted that cybercriminals already use publicly available and customized AI tools to further their deceptive activities, making them even more challenging to detect. "As technology continues to evolve, so do cybercriminals' tactics. Attackers are leveraging AI to craft highly convincing voice or video messages and emails to enable fraud schemes against individuals and businesses alike," says FBI special agent in charge Robert Tripp. "These sophisticated tactics can result in devastating financial losses, reputational damage, and compromise of sensitive data."[6]

CHAPTER ONE

# What Is Artificial Intelligence?

Artificial intelligence is a powerful tool used everywhere, from schools and businesses to hospitals and government agencies. Even though most people have heard of AI, many are confused about what it is. AI is a type of computer software based on the human brain and how it learns. AI software teaches computers to think and learn like humans. It works by analyzing large amounts of data, learning from it, and improving over time.

Using AI, computers can solve problems, recognize patterns, and make decisions that typically require human intelligence. Some types of AI can recognize speech, translate languages, or even create images and text. AI powers applications such as smartphone voice assistants, website chatbots, self-driving cars, and streaming video recommendation systems. While AI makes life easier in many ways, the technology also raises concerns about how it can be misused for deception.

## Everyday AI Examples

AI has made many everyday tasks more efficient and effective. For example, when a person asks a question to a voice assistant such as Siri or Alexa, the technology uses AI to understand the spoken language and respond to commands. AI-powered voice assistants can efficiently perform tasks such as setting reminders, giving weather updates, playing music, and controlling smart home devices.

AI personalizes the advertisements and content individuals see online as they scroll through social media feeds or browse websites. AI algorithms analyze a person's activity online, gathering data about what sites they visited, what they liked on social media, and what content they interacted with. AI algorithms use this data to personalize the posts, ads, and recommendations people see each time they go online.

> **"AI also paves the way towards unimaginable applications that can better our day-to-day life by enhancing how we communicate, conduct business, and navigate the world."[7]**
>
> —Walid Saad, computer engineering professor at Virginia Tech

Video games use AI to adapt gameplay to a person's actions and performance. AI can improve the gaming experience by creating personalized challenges, adjusting difficulty levels, and creating scenarios that match an individual player's style and preferences.

In health care, AI technology helps doctors diagnose and treat patients. AI algorithms analyze medical images, such as X-rays, computed tomography scans, and magnetic resonance imaging. AI technology quickly searches for abnormalities or other details that allow doctors to diagnose patients accurately. AI technology can also analyze patient data and predict how a disease will progress. The technology can use this information to recommend a personalized treatment plan, which can improve patient care.

The possibilities appear limitless for AI applications. Walid Saad, a computer engineering professor at Virginia Tech, says:

> AI also paves the way towards unimaginable applications that can better our day-to-day life by enhancing how we communicate, conduct business, and navigate the world. Sectors that can benefit from automation range from transportation to healthcare, telecommunications, agriculture, production, and even entertainment. Technologies that we take for granted now, like home assistants, recommendation systems (e.g., those we see in YouTube or Netflix), or robotics, could not have been possible without AI, and we will continue to see them evolve further as the power of AI improves.[7]

## Creating and Training AI Tools

Designers create AI tools by writing computer code that creates a model that receives and learns from data inputs. The model trains on enormous amounts of data, which it analyzes for patterns and correlations. Then the model uses these patterns to create rules called algorithms. The algorithms give the computer step-by-step instructions to perform a specific task and make predictions or decisions based on the data. Over time, the AI model learns to choose the correct algorithm to perform the desired task and produce the desired output.

AI systems can be built and trained to perform a wide range of tasks, from recognizing images to translating text from one language to another. The data used to train an AI model varies depending on the model's purpose. For example, an AI image-recognition tool will train with millions of images so that it can learn to identify objects in images. This type of AI tool could be used to scan medical images and identify specific conditions. An AI chatbot will train on customer data and analyze it so that it can provide accurate and personalized responses. This type of

***AI powers technology such as this self-driving car and many other applications. While AI makes life easier, the technology also raises concerns about how it can be misused for deception.***

chatbot could be used as a recommendation tool for a music or movie streaming service.

Depending on the model's size and amount of training data, AI training can last for weeks or even months. Some models train on datasets that hold billions of words or images that are input and analyzed. The more training the AI model completes, the faster, more precise, and more effective it can be. The more data a model analyzes, the better its algorithms and results become.

> **"These tools can make your life and your job easier and, as a result, enable you to do more meaningful work."[8]**
>
> —Georgia Perakis, professor at the MIT Sloan School of Management

Once trained, AI technologies are powerful at information processing, turning data into useful information. AI can process and analyze massive amounts of data quickly and accurately, much faster than humans. These skills make AI systems good at tasks that require recognizing images, translating language, and predicting trends. "AI models allow us to process large amounts of data, run an algorithm, and run it fast. As a human being, you cannot process the same amount of data and determine optimal strategies in complex settings. But you have the experience to interpret it and use it with care. These tools can make your life and your job easier and, as a result, enable you to do more meaningful work,"[8] says Georgia Perakis, a professor at the MIT Sloan School of Management.

## Types of AI

Some types of AI, known as narrow or weak AI, are built to perform a specific task. Narrow AI is particularly good at the task it was built to do. However, narrow AI cannot learn new tasks outside its original purpose. For example, Netflix's recommendation engine excels at making movie recommendations but cannot answer a question about the weather.

Google Translate, a language translation app, is an example of a narrow AI system. Google Translate was specifically designed to translate text from one language to another. A user enters text or speaks into the app in one language, and the app translates it into another language. The AI system processes and analyzes the input data using algorithms trained on language data from the source and output languages. Then the AI system produces a translation in the chosen language. Google Translate is good at what it was trained to do, translating language, but it does not create anything new.

In contrast, another type of AI, known as generative AI, creates original material. Generative AI systems create new text, images, music, videos, computer code, and more. While traditional AI models use training data to learn how to analyze data and make predictions, generative AI models use training data to learn how to create entirely new data. ChatGPT is an example of a generative AI system. ChatGPT's model is trained on enormous amounts of text so that it can generate original text. Users enter a prompt, and ChatGPT generates an original response almost identical to human-written text.

Generative AI can be deployed to make many tasks more efficient. Generative AI can help write reports, emails, and other documents, summarize large amounts of text, and generate images from a prompt. It can generate natural-sounding speech for chat-

*Netflix's recommendation engine is an example of narrow AI. The engine excels at making movie recommendations, but it cannot learn new tasks outside of its original purpose.*

## Computer Vision

Computer vision is a branch of artificial intelligence that enables computers to process, analyze, and understand visual data from the world, similar to how humans perceive images and videos. Computer vision AI extracts meaningful information from digital images or videos and makes decisions based on that data. Computer vision allows machines to detect, classify, and track objects. It also teaches machines to recognize faces and interpret visual scenes.

Computer vision can be used in various applications. In self-driving cars, computer vision helps detect pedestrians, traffic signs, and other obstacles so the car can drive safely. In health care, computer vision helps doctors read medical images and diagnose diseases. Security and surveillance tools can also use computer vision to scan visual scenes and give an alert when abnormalities are detected. In manufacturing, computer vision can scan thousands of products and learn to quickly spot defects before products are shipped to customers.

bots and other applications. Generative AI can make the design process more efficient by creating many prototypes quickly. In the entertainment industry, generative AI can create new music, write scripts, or even create videos. It has the potential to make work easier in fields where creation is essential.

## Machine Learning

Machine learning is a type of AI that enables computers to learn and improve from data without being specifically programmed. Machine learning relies on algorithms written by humans. Algorithms give computers step-by-step directions, showing them how to analyze and learn from data. AI uses algorithms to process information, find patterns, and make decisions.

An email spam filter is an example of machine learning AI. Unwanted spam emails are a nuisance for many users and can even be harmful if they contain malware or phishing scams. Email service providers use AI-powered spam filters to catch spam emails. Designers gather a large amount of training data, which are emails labeled as spam or not spam. A machine learning algorithm is trained using this data. As it trains, the algorithm learns to recognize patterns in the email's features and their labels as spam or not spam. The algorithm is then used in a model that can accurately classify new emails based on the patterns learned in training.

After the model is trained and tested, the spam filter goes to work. As each new email is received, the spam filter analyzes it and decides whether it is spam, using what it has learned from past emails. Emails with a high probability of being spam are sent to the email's spam folder. The spam filter learns from its mistakes. If an email is sent to the spam folder in error, users can mark it as not spam and return it to their inbox. The spam filter uses that information to perform better the next time.

## Neural Networks and Deep Learning

Another type of AI model, known as an artificial neural network, mimics the way the human brain works. The human brain is made up of cells called neurons, which are connected and send signals to each other. The brain's neurons are connected across a netlike structure and use electrical impulses to send data between neurons. These connections enable a person to store and process information and perform complex tasks. An artificial neural network is a digital version of the brain's neural network.

An artificial neural network is a computer system made from interconnected nodes like the brain's neurons. The nodes work together to process information. Each node receives input data, processes it, and then sends the output to other nodes. An artificial neural network is like a group of people sharing and processing data together. Each person sends messages to others in the network. Some people are good at noticing specific details in the data, while others are better at recognizing patterns. Working together and sharing messages, the group can solve many types of problems.

Artificial neural networks use layers to process and analyze data. Each layer of nodes examines data in a different way. The first layer of nodes may look for basic information, while subsequent layers search for more complex details. For example, an AI

*Smartphone voice recognition uses deep learning. The artificial neural network's layers analyze speech patterns, phonetics, and other characteristics to understand speech.*

system used for facial recognition might use artificial neural networks and layers. The first layer might identify basic facial features such as colors and edges. As the data moves from layer to layer in the network, each layer searches for and identifies more complex details, such as shapes and textures. Eventually, the network can identify specific features such as eyes and noses. Breaking down the process into layers, artificial neural networks are good at analyzing and learning complex patterns.

Deep learning is an application of AI that uses artificial neural networks with several layers to process complex data. Smartphone voice recognition uses deep learning. The artificial neural network's layers analyze speech patterns, phonetics, and other characteristics to understand speech. The smartphone can understand and respond to spoken commands accurately. Other AI systems that use deep learning include self-driving cars and medical image diagnostic tools. Ella Atkins, an engineering professor at Virginia Tech, says:

> Machine learning (ML) in the form of deep neural networks has unquestionably revolutionized our ability to process large datasets and classify or otherwise learn from their

content. For example, in the health space, medical issues such as cancer detection are now significantly enhanced by ML perception. These types of societal impacts will continue as artificial intelligence and machine learning continue to grow and evolve.[9]

## Natural Language Processing

Natural language processing (NLP) is a type of AI that teaches computers to understand, interpret, and generate human language. NLP algorithms process and analyze text to learn how words and sentences are formed, what they mean, and how people use them. The more text they process, the better NLP algorithms get. They learn how people talk so they can better understand and respond. This learning enables computers to write better text, translate languages instantly, or chat like humans. Text-to-speech apps are one example of AI systems that use NLP.

### Learning to Deceive

Technology experts warn that some AI systems are learning how to deceive humans even when they were not trained to do so. For example, CICERO is an AI system developed by Meta in 2023 to play Diplomacy, a world conquest strategy game. CICERO proved to be a good player and ranked in the top 10 percent of human players. In training, CICERO communicated in text with human players and learned to form alliances. The AI system also learned how to deceive and betray alliances when doing so benefited its game. Other game-playing AI systems have also demonstrated deception, such as bluffing during a poker game or faking attacks during a strategy game to defeat opponents. "AI developers do not have a confident understanding of what causes undesirable AI behaviors like deception," says Peter Park, a Massachusetts Institute of Technology postdoctoral fellow who studies AI. "But generally speaking, we think AI deception arises because a deception-based strategy turned out to be the best way to perform well at the given AI's training task. Deception helps them achieve their goals." Cheating in games might appear harmless, but Park warns it could lead to more advanced AI deception in the future.

Quoted in Cell Press, "AI Systems Are Already Skilled at Deceiving and Manipulating Humans," ScienceDaily, May 10, 2024. www.sciencedaily.com.

> **"While the transformative potential of AI is undeniable, so too are the multifaceted risks it presents."[11]**
>
> —Rohit Kundu, PhD student at the University of California, Riverside

A website chatbot is an AI-powered tool that uses NLP to help it generate human conversation, which allows it to talk to users via text. When a user types a message, the chatbot figures out the user's words, determines the meaning, and then chooses the correct response based on a set of rules or from learning. As NLP advances, chatbots get better in their responses.

Large language models are a subset of NLP that focuses on generating text. They make it easier for humans to communicate with machines. "Large language models (LLMs) are transforming our interactions with technologies. Their capacity to parse and generate human-like text has made it possible to have more dynamic conversations with machines. These models are no longer just about automating tasks—they are versatile support tools that people can tap into for brainstorming, practicing tough conversations, or even seeking emotional support,"[10] says Eugenia Rho, assistant professor of computer science at Virginia Tech.

## Potential for Misuse

AI is a powerful technology that can make many tasks easier and more efficient. Yet in the wrong hands, AI has the potential for misuse and deception. The characteristics that make AI useful in everyday life also make it dangerous when misused. AI-generated text, images, video, and audio can appear to be from real people. These can be used to commit plagiarism, spread misinformation, manipulate people, and carry out scams and fraud. "Artificial intelligence (AI) has swiftly transitioned from a futuristic concept to an integral part of everyday life. From virtual assistants to recommendation algorithms, AI has become the new normal, permeating industries and revolutionizing how we interact with technology. While the transformative potential of AI is undeniable, so too are the multifaceted risks it presents,"[11] says Rohit Kundu, a University of California, Riverside, PhD student whose focus is deep learning.

# Plagiarism and Cheating

During Hannah's first year at a university in Britain, the pressure to succeed was intense. When she got sick with COVID-19, the assignments piled up. "I felt incredibly stressed and just under enormous pressure to do well. I was really struggling and my brain had completely given up,"[12] she says. Facing two back-to-back deadlines, Hannah turned to artificial intelligence to help her write one of her essays. Using AI to complete her work violated her school's code of conduct.

Hannah's use of AI to write her essay was discovered when her professor ran the assignment through AI-detection software, which flagged Hannah's essay as being written by AI. The professor gave Hannah a zero on the assignment and referred her case to the university's academic misconduct panel. Hannah admitted to the panel that she had used AI to help her write the essay. She was relieved when the panel decided not to expel her.

Hannah believes her experience can serve as a warning to other students about the risks of using AI to cheat. "I could have been kicked out," she says. "I do massively regret my choice. I was achieving really well . . . and I actually think that might have also been the problem, that I needed to maintain that level of grades, and it just kind of really pushed me into a place of using artificial intelligence. . . . It felt really bad at the end of it, it really tainted that year for me."[13]

## AI Plagiarism on Campus

Some students use AI to generate ideas for an upcoming report or paper. This type of AI use is generally considered appropriate by many teachers and schools. The problem arises when students go further and use AI to write the paper. Using AI tools and passing the work off as one's own is a type of cheating called plagiarism. Plagiarism is the act of using someone else's work or ideas without giving them credit. Plagiarism is a serious academic offense and can have severe consequences. Students caught turning in plagiarized work may be given a failing grade, receive academic probation, or even get expelled.

AI plagiarism occurs when individuals submit text generated by an AI program as their original work without citing the AI source. They pass off the AI-written content as if they wrote it themselves, which is considered academic plagiarism. For example, a student might use an AI tool like ChatGPT to write a large part of an essay or paper and then submit it without noting that he or she used AI to generate a good portion of it. Maddy Osman, an online content creator, says:

> As humans, we often perform research and use the work of other creators to inform our own, ideally with proper credit and citation. The difference is that each human processes and interprets their research differently and uses it to create something new. When an AI references human work to generate content, that same level of cognitive processing doesn't happen. In addition to lacking citations, AI-generated content is simply a regurgitation of the data it's been trained on without adding a unique and individual perspective.[14]

Darren Hick, a philosophy professor at Furman University, first noticed an AI-generated essay from one of his students in 2022. By 2024, Hick observes, AI use had spread like a virus among

> "All plagiarism has become AI plagiarism at this point. I look back at the sort of assignments that I give in my classes and realize just how ripe they are for AI plagiarism."[15]
>
> —Darren Hick, philosophy professor at Furman University

students on his campus. "All plagiarism has become AI plagiarism at this point. I look back at the sort of assignments that I give in my classes and realize just how ripe they are for AI plagiarism,"[15] he says.

Hick's experience with students using AI to write papers is not unique. In 2023 software company Turnitin developed an AI-detection tool that could be used to identify AI-generated writing. Since its release, the plagiarism checker has reviewed over 200 million papers, written mainly by high school and college students. The tool detected more than 22 million papers that were at least 20 percent written by generative AI. Approximately 6 million of those AI-generated papers were flagged for being 80 percent or more AI-generated writing.

*AI plagiarism occurs when individuals submit text generated by an AI program as their original work without citing the AI source.*

## Challenges with AI-Detection Tools

However, Turnitin and other similar AI-detection tools are not infallible. When analyzing an entire document, the company says that its AI-detection tool has a false positive rate of less than 1 percent. The false positive rate is slightly higher, about 4 percent, for partial documents or individual sentences. A false positive flags writing for using generative AI when the author did not do so. Because of the potential for false positives, Annie Chechitelli, Turnitin's chief product officer, recommends that professors and teachers talk to students before accusing them of AI plagiarism. "It's just supposed to be information for the educator to decide what they want to do with it. It is not perfect,"[16] says Chechitelli.

Liberty University senior Maggie Seabolt knows what it feels like to be falsely accused of AI plagiarism. A professor flagged Seabolt's paper in one class for having 35 percent AI-written content. Seabolt was confused because she knew she had written the paper in Microsoft Word in one sitting. "To see that I was being accused of using AI when I knew in my heart I didn't, it was really, really stressful, because I had no idea how to even prove my innocence. I definitely felt very alone,"[17] says Seabolt. The professor marked Seabolt's paper grade down by 20 percent because of the AI accusation.

New research also suggests that AI detectors are not always reliable. A 2024 study by University of Pennsylvania researchers found that AI detectors can be easily fooled. Simple tricks such as adding white space to text, misspelling words, removing grammar markings, and using characters called homoglyphs that look like letters and numbers caused many AI detectors to fail. "That breaks these AI detectors, and their performance drops by like 30%,"[18] says study author Chris Callison-Burch. Researchers also found that AI detectors struggled to identify content created by lesser-known AI models.

The University of Pennsylvania study also found that many AI detectors set their models to allow high false positive rates. A high false positive rate is more likely to detect AI-generated content,

*A recent study showed that AI detectors, like Turnitin, can be easily fooled. Simple tricks such as misspelling words and removing grammar markings caused many AI detectors to fail.*

but it is also more likely to incorrectly label content written by a person as AI-generated. Researchers found that accuracy levels dropped when these AI detector models used a more reasonable false positive rate. "These claims of accuracy are not particularly relevant by themselves. . . . I would use these systems very judiciously if you're a professor who wants to forbid AI writing in your classrooms. Probably don't fail a student for using AI just based on evidence of these systems, but maybe use it as a conversation starter,"[19] says Callison-Burch.

## Cheating on Homework and Tests

Students also use AI to do homework and complete take-home tests and quizzes. For example, students sometimes take pictures of math problems and upload them to an AI tool that gives them step-by-step instructions on how to solve the problems. While AI can be a useful tool to help students understand a difficult concept such as how to solve a math problem, using AI for a test or to complete an assignment is not ethical when the purpose of the test or assignment is to show what the student

> **"Generative AI can be a great tool to boost productivity, but unfortunately, many people, especially teens, are seeing it as a shortcut."[20]**
>
> —Jack Kosakowski, president of Junior Achievement USA

has learned. In a 2023 survey by Junior Achievement USA, more than 44 percent of teens said they would likely use AI to complete school assignments instead of doing the work themselves. Nearly half said they knew a peer who had already used AI to complete school assignments. And in the same survey, a majority of teen respondents (60 percent) said they believed using AI for school assignments was cheating.

Others are concerned that relying on AI to complete schoolwork prevents students from thoroughly learning the content. Jack Kosakowski, president of Junior Achievement USA, says:

> Generative AI can be a great tool to boost productivity, but unfortunately, many people, especially teens, are seeing it as a shortcut. The misuse of AI to do all schoolwork not only raises ethical concerns, but this behavior could also short-change many students' educations since they may not be learning the subjects they are using AI for. Given the growing demand for marketable skills, this could become very problematic.[20]

## Adapting Assignments

Some teachers and professors are changing the type of assignments they give students to reduce and prevent AI plagiarism and cheating. Instead of assigning a traditional essay, Helena Kashleva, an adjunct instructor at Florida SouthWestern State College, plans to assign critical thinking exercises or personalized reflections that are difficult to complete with AI. Janine Holc, a professor of political science at Loyola University Maryland, is assigning essays to be handwritten and completed during class time, preventing students from turning to online AI tools. Some teachers have shifted to assigning more oral presentations, classroom debates, and group projects to reduce student reliance on AI tools. Others require students to turn in rough drafts, outlines, and research notes to show their progress on written work. Some, like Kerry O'Grady, an associate professor of public relations at Columbia University, are incorporating AI into their lesson plans, teaching students how to use AI tools ethically and highlighting the potential errors AI can produce.

> "A geography assignment was due next period, and I used ChatGPT to write out the whole speech for me. When I was saying it out loud and I got asked questions, I had no clue what I was saying. I got a detention."[21]
>
> —British teen caught using AI for schoolwork

When using AI to do homework violates school rules, students risk punishment if caught. One British student shared his experience of using AI for homework. "A geography assignment was due next period, and I used ChatGPT to write out the whole speech for me. When I was saying it out loud and I got asked questions, I had no clue what I was saying. I got a detention,"[21] he says.

## AI and Scientific Shortcuts

Beyond the classroom, generative AI has been problematic in the scientific research community. Researchers who work for universities often have significant pressure to publish papers in academic journals. Publishing, or the lack of it, is one of the main ways a scientist is evaluated for a promotion or new job. The publishing mindset is so common that "publish or perish" is a common phrase in the community.

Publishing a research paper in a reputable academic journal is often a long, involved process. After months or years of research, the paper is submitted for peer review by other scientists, which

Students sometimes take pictures of math problems and upload them to an AI tool that gives them step-by-step instructions on how to solve the problems.

## AI Image Generators

Some researchers have used AI to generate images in scientific papers, a practice that has sparked debate. Supporters point out that AI-generated images are useful for visualizing complex concepts. AI-generated images can help simulate models, create visuals of complicated theories, and improve educational materials. However, critics argue that AI-generated images can be misused to falsify or manipulate scientific data and mislead readers and reviewers. Critics also point out that manipulated images can lead to false conclusions and questions about the reliability of scientific data. Many scientific journals and institutions have implemented strict guidelines about using AI-generated images. They frequently require authors to disclose when they use AI and, in some cases, require authors to provide the raw data used to generate the images.

typically leads to several revisions. Sometimes, peer reviewers and journal editors ask the scientist to perform additional experiments and add the results to their paper. This process, from paper submission to publication, is a lengthy one. Most of the time, it results in an authenticated and polished scientific paper.

When used responsibly, AI has the potential to be a useful tool for scientific research and papers. AI can help researchers gather and analyze massive amounts of data. It can also translate from one language to another. In this way, AI tools can lower language barriers, accelerate publication times, and increase efficiency. However, some researchers looking to cut corners in publishing have turned to generative AI to help them write research papers. In 2023 a survey of scientists by the well-known journal *Nature* revealed that about 30 percent admitted they had used AI tools to help them do just that. When researchers use AI to generate academic papers without disclosing its use, it is a form of academic dishonesty. While academic dishonesty in science existed long before AI, the widespread availability of generative AI, such as ChatGPT, has caused it to become more common in scientific papers. When scientists take shortcuts with AI tools in scientific papers, it can lead to errors, inaccuracies, and made-up results, which can threaten a journal's reputation. As a result, in 2023 academic journals retracted nearly fourteen thousand papers, compared to about two thousand retractions in 2013, according to the *Scientist*. Most of the retractions involved plagiarism, academic dishonesty, and concerns about authentic data.

Many journals do not agree on whether to allow AI-generated text in their published papers. For example, journals published by *Science* strictly ban all AI-generated text or images if not first approved by an editor. *Nature* bans AI-generated images and video but allows AI-generated text in some instances. *JAMA* allows AI-generated text but requires it to be disclosed. The variation in AI policies creates confusion for scientists submitting papers and the reviewers who read them.

Generative AI tools have the ability to create new content, including text and images. While this technology can make many tasks more efficient, concerns arise about potential misuse and deception. Students and professionals may misuse AI to produce essays, articles, reports, and more without giving proper citations. Detecting improper AI use remains complicated as AI-detection tools have challenges with accuracy and false positive rates.

# Misinformation and Manipulation

In 2022 CNET, a website that publishes news and reviews about consumer electronics and technology, quietly began using AI to write articles published on its site. Dozens of articles written by "CNET Money Staff" were entirely AI generated. Readers only discovered the article's AI source if they clicked a link in the staff byline. Problems began to emerge when a closer look at some of CNET's AI-generated articles uncovered factual errors. For example, an AI-generated article about compound interest stated that a $10,000 deposit earning 3 percent interest would earn $10,300 in the first year. The interest calculation was inaccurate. In reality, the deposit would only earn $300.

When notified about the AI errors, CNET paused the publication of all AI-generated articles and launched a review of all AI-generated content on its site. In an editorial, CNET's editor in chief, Connie Guglielmo, explained the company's use of AI to generate content. Guglielmo explained that CNET had used an internally designed AI tool to help craft seventy-seven published articles during November 2022 to January 2023. She explained that it was part of a test project for the CNET Money team. "Editors generated the outlines for the stories first, then expanded, added to, and edited the AI drafts before publishing," Guglielmo said. "After one of the AI assisted stories was cited, rightly, for factual errors, the CNET Money editorial team did a full audit."[22]

The audit of CNET's AI-generated articles revealed that the problems reached beyond the original incorrect articles. The audit identified multiple stories with errors that needed corrections.

A small number of these stories required significant revision. CNET also identified other problems with the AI-generated articles, such as incomplete company names, transposed numbers, or vague language.

CNET had initially stated that its AI-generated articles had been edited and fact-checked by a human editor. Hany Farid, professor of computer science at the University of California, Berkeley, suspects the human editors over-relied on AI technology. "I wonder if the seemingly authoritative AI voice led to the editors lowering their guard and [being] less careful than they may have been with a human journalist's writing,"[23] he says.

## AI in the Newsroom

CNET is not the only news outlet that has been experimenting with AI to research and write content. The prestigious news agency Associated Press (AP) began using an AI system to automatically generate quarterly earnings stories as far back as 2014. Within six months, the AP had ramped up to publishing three thousand AI-generated stories every quarter. However, the AP's use of AI was fairly simple, since the technology inserted earnings information into preformatted article templates.

As AI technology advances, AI systems can provide several benefits for news organizations. AI systems are quick and efficient, and when used to automate mundane tasks, they can free journalists to focus on investigations and more complex reporting. AI systems can quickly analyze large datasets, go through many documents, and summarize reports. AI technology helps journalists search through news archives to see previous articles on a topic and identify quote sources. AI systems can generate news articles in real time, enabling news outlets to deliver news faster.

## Risk of Misinformation and Bias

Problems arise, however, when news outlets rely on AI to generate news content with little human oversight and disclosure to the

In 2022 CNET, a technology news website, quietly began using AI to write articles published on its site. It was discovered that many of the articles had factual errors.

public. AI systems are only as good as the algorithms that power them and the data they use for training. Where there are problems with the underlying algorithms or training data, AI-generated content can contain unintended errors, misinformation, and bias.

Sometimes, AI systems generate inaccurate information, a phenomenon called a hallucination. AI hallucinations can be minor, such as stating the wrong historical date or getting a person's first name wrong. Other times, AI errors are more serious, such as when an AI-generated article gives the wrong health or financial advice. AI hallucinations can occur in text generated by large language models (LLMs) and in AI-generated images.

Google's new AI chatbot, an LLM originally called Bard, demonstrated the risk of hallucinations firsthand in 2023. In a promotional video that same year, Bard was asked about new discoveries from the James Webb Space Telescope (JWST). Bard incorrectly answered that the JWST was used to take the first pictures of a planet outside Earth's solar system. Experts quickly pointed out that Bard was wrong. "I'm sure Bard will be impressive, but for the record: JWST did not take 'the very first image

## Zero Tolerance for AI

The Arizona State University student newspaper, the *State Press*, has a zero tolerance policy for AI-generated content because of concerns about academic integrity and unintentional misinformation on its platform. So, in 2024 when editors discovered that a student journalist had used AI to write weekly horoscopes, news articles, and editorials, they fired the journalist. The paper's editors reviewed all of the journalist's work and ended up retracting twenty-four articles because they were at least partially written with AI. The paper also verified the accuracy of the articles, removed original bylines, and replaced the text of several articles with a notice explaining that generative AI had been used to produce the original content.

of a planet outside our solar system,'"[24] says Grant Tremblay, an astrophysicist at the Center for Astrophysics. That first image was taken in 2004 by the Very Large Telescope (VLT) at the European Southern Observatory in northern Chile.

Like other LLMs, Bard (renamed Gemini in 2024) trains on massive amounts of text data from the internet. In this way, the AI model learns how to respond to text prompts. However, when there were errors in the training data, Bard repeated them. "This highlights the importance of a rigorous testing process, something that we're kicking off this week with our trusted tester program. We'll combine external feedback with our own internal testing to make sure Bard's responses meet a high bar for quality, safety, and roundedness in real-world information,"[25] said a Google representative at the time.

AI algorithms can also be affected by bias. A model that trains on biased data will generate biased content. Francisco Castro, a UCLA Anderson School of Management professor, explains how bias can be introduced into AI models. "When I'm programming my AI, let's say I only use data from the *New York Times*, or maybe I only use data from Fox News. Then my model is only going to be able to generate output from that data," he says. "It's going to generate a biased output that doesn't necessarily represent the [diversity] of opinions that we observe in the population."[26]

## Intentional Manipulation

Although some of the errors, misinformation, and bias in AI-generated content are unintentional, some are not. AI can be used to generate news and other content that is designed to mislead and manipulate readers and viewers. Once these articles are ready for public consumption, they are released on social media platforms and other online sites.

Manipulative websites that exist primarily to spread false and misleading information are not new. However, the emergence of AI has given bad actors a powerful tool that makes their job easier. "With the advent of AI, it became easier to sift through large amounts of information and create 'believable' stories and articles," says Virginia Tech's Walid Saad. "Specifically, LLMs made it more accessible for bad actors to generate

*Google's AI chatbot Gemini (originally named Bard) trains on massive amounts of data. If errors are present in the data, Gemini repeats them rather than corrects them.*

what appears to be accurate information. This AI-assisted refinement of how the information is presented makes such fake sites more dangerous."[27]

The number of websites hosting AI-generated misinformation is growing quickly. In May 2023 NewsGuard, an organization that tracks online misinformation, identified forty-nine websites in a single month that appeared to be generated entirely or mainly by AI systems. By December 2023, NewsGuard had identified more than six hundred sites that hosted AI-generated false articles. The rapid increase of these sites is also due to the power of AI. In the past, creating false content and building sites that appeared legitimate to host false content required dozens of low-paid workers. With AI, almost anyone with a computer and internet access can create these websites and publish content that is difficult to distinguish from real news.

> **"Some of these sites are generating hundreds if not thousands of articles a day. This is why we call it the next great misinformation superspreader."[28]**
>
> —Jack Brewster, researcher at NewsGuard

AI-generated content has made it easier than ever to mislead people online. AI programs, particularly LLMs, generate text that sounds like a real person wrote it. AI video tools create realistic videos based on text prompts that look like professionally produced clips. Also, the sheer volume of false AI content makes it hard to know what is true and what is fake. Some sites publish hundreds of articles daily, many containing errors or promoting misinformation. "Some of these sites are generating hundreds if not thousands of articles a day," says Jack Brewster, a researcher at NewsGuard. "This is why we call it the next great misinformation superspreader."[28] The problem can multiply when artificial intelligence LLMs pull false content for training data and thus create a cycle of errors and misinformation.

## Clickbait and Disinformation

Sometimes, AI-generated content is created as clickbait, intentionally designed to increase user clicks and engagement. Online

## *Sports Illustrated* AI Scandal

In 2023 reporters at Futurism, a technology news company, claimed that *Sports Illustrated* had published AI-generated articles written by fake authors. Article bylines listed human writers who did not exist and had AI-generated headshots and biographies. At first, *Sports Illustrated* denied the claims. The magazine explained that the content was created by a third-party company, AdVon Commerce, that had been hired to produce website content. While *Sports Illustrated* insisted the content itself was not AI-generated, the magazine removed the controversial content from its website and ended its agreement with AdVon Commerce. However, the damage to the magazine's reputation had already been done.

While many people believe that media companies like *Sports Illustrated* should experiment with AI, they want the media to be honest and up-front about it. "The mistake is in trying to hide it, and in doing it poorly," says Tom Rosenstiel, a University of Maryland professor who teaches journalism ethics. "If you want to be in the truth-telling business, which journalists claim they do, you shouldn't tell lies. A secret is a form of lying."

Quoted in David Bauder, "*Sports Illustrated* Found Publishing AI Generated Stories, Photos and Authors," *PBS NewsHour*, November 29, 2023. www.pbs.org.

advertisers typically pay website owners for clicks. The more users who click on links and engage with a website's content, even if that content is false, the more money the site's owners can make in advertising revenue.

Some bad actors have more in mind than making money from AI-generated content. They purposely use AI-generated content to deceive and manipulate public opinion. For example, Russia launched a sophisticated influence campaign known as DoppelGänger to influence how the world viewed the country after it invaded Ukraine in February 2022, according to cyber experts at the European Union DisinfoLab. Going back to at least May 2022, Russian agencies had created a network of cloned websites, fake articles, and social media posts designed to spread disinformation in Europe and beyond.

To carry out the DoppelGänger campaign, Russian representatives bought domain names online that were similar to those of real media outlets. They pretended these sites belonged to legitimate new agencies, governments, and think tanks. Then they used generative AI systems to create and publish disinformation on these sites to trick unsuspecting readers into believing and spreading pro-Russian views. For example, DoppelGänger set up a domain similar to the official North Atlantic Treaty Organization (NATO) website and published AI-generated press releases that made several false claims. Some fake press releases claimed that NATO countries were debating sending Ukrainian paramilitary troops to France to suppress protests. Another website published pro-Russian false content that even claimed to have undergone fact-checking. Some articles were false pro-Russian stories about the harms caused by Ukrainian refugees in Europe, while others falsely claimed that Western economic sanctions against Russia were ineffective.

Next the DoppelGänger campaign employed social media chatbots to spread its pro-Russian disinformation further online.

*This site in Ukraine was destroyed by Russian troops in 2022. Russia is accused of posting AI-generated false information about its war efforts in Ukraine.*

"This cross-platform campaign amplifies the deceptive content distributed through its cloned web pages across various social media networks, including Facebook and Twitter. Videos, articles, and polls designed to manipulate public opinion are disseminated seamlessly, blurring the lines between fact and fiction,"[29] warned US Cyber Command in a September 2024 statement.

## Spreading on Social Media

Once AI-generated disinformation hits social media, it spreads easily. Many social media users share content without verifying the origin or accuracy of the information. Sometimes, the social media platform's algorithms boost AI-generated posts that are filled with false and misleading information—ensuring that more people see and share them.

In 2024 researchers at Georgetown University and Stanford University investigated over one hundred Facebook pages that regularly posted AI-generated content. Researchers found that Facebook's algorithms promoted many AI-generated posts that were nothing more than clickbait scams. First, Facebook's algorithms actively pushed the posts onto users' feeds. Then, as users engaged with the posts, Facebook's algorithms recommended them to more users. Some of the AI scams investigated had millions of user interactions.

The motivations behind these AI scams were not always clear. Some were clearly intended to generate revenue, while others appeared to be designed to gain more followers. "It could be that these were nefarious pages that were trying to build an audience and would later pivot to trying to sell goods or link to ad-laden websites or maybe even change their topics to something political altogether,"[30] says Josh Goldstein, a research fellow at Georgetown University and coauthor of the study.

Many people are worried about the impact of AI-generated disinformation and false images on social media. "It just sort of reinforces people's disbelief and . . . makes it harder to see what is real," says Hobey Ford, a social media user in North Carolina. Ford

> **"The danger is the scope and scale with AI . . . especially when paired with more sophisticated algorithms. It's an information war on a scale we haven't seen before."[32]**
>
> —Jeffrey Blevins, misinformation expert and journalism professor at the University of Cincinnati

says that he has seen AI images in Facebook science groups that claim to show new discoveries. "And I think that's dangerous in our world right now,"[31] he says.

Advances in generative AI have enabled anyone with a computer and internet access to create and publish misleading and manipulative content on websites and social media platforms. As AI technology advances, it has become more challenging to determine what is real and what is deceptive. "The danger is the scope and scale with AI . . . especially when paired with more sophisticated algorithms. It's an information war on a scale we haven't seen before,"[32] says Jeffrey Blevins, a misinformation expert and journalism professor at the University of Cincinnati.

# Deepfakes

New York attorney general Letitia James issued an investor alert in August 2024. She warned New Yorkers about scams promoting investment in cryptocurrencies and other financial products. The scams used AI technology to create videos that appeared online, often in social media feeds, digital ads, and messaging apps. The AI videos, called deepfakes, featured well-known, wealthy individuals such as Warren Buffet, Jeff Bezos, and Elon Musk, encouraging viewers to invest in scams often involving cryptocurrency. "Sophisticated scammers are using AI to impersonate trusted business leaders and scam vulnerable New Yorkers out of their hard-earned money. Manipulated videos advertising phony investment scams are spreading like wildfire on social media, and New Yorkers should know how to avoid falling victim to these schemes,"[33] James said.

> **"Sophisticated scammers are using AI to impersonate trusted business leaders and scam vulnerable New Yorkers out of their hard-earned money."[33]**
>
> —Letitia James, New York attorney general

To create the videos, scammers used AI tools to manipulate an existing video of a real person. They altered the person's mouth movements and speech so the person appeared to endorse the scam. The scammers then posted the deepfake videos online as social media posts, fake live streams, and online advertisements. When someone showed interest in the investment, the scammers convinced the victim to move further conversations to private message services such as WhatsApp or Signal. Once the victim invested, the scammers directed him or her to a fake website showing fake profits. Often, this allowed the scammers to con victims into sending more money for additional investments, sometimes as much as hundreds of thousands of dollars. However, if the victims attempted to withdraw their funds, they could not do so or were told they would need to pay taxes on their withdrawals. Eventually, the scammers shut down the fraudulent websites and disappeared, taking the victims' money with them.

Heidi Swan, a sixty-two-year-old health care worker, was one of the New Yorkers swindled by an AI deepfake investment scam. Swan saw ads for a cryptocurrency investment on several social media sites, including Facebook and TikTok. In the ad, billionaire Elon Musk appeared to pitch the investment opportunity. If a billionaire businessman like Musk supported the opportunity, Swan thought the investment must be a good idea. "Looked just like Elon Musk, sounded just like Elon Musk, and I thought it was him,"[34] says Swan. She contacted the company in the ad and opened an investment account with $10,000. However, Swan later learned that she had been scammed by bad actors who used AI to create the deepfake video of Musk.

## Making Deepfakes

Fake photos or altered images are common online. Often, these alterations are harmless, such as when a photo is altered by switching two people's faces or individuals use an aging app to create a photo of what they will look like in the future. Not only are these uses harmless, they are usually easy to spot.

AI-generated deepfakes are much more difficult to detect, which makes them more dangerous. Deepfakes are a type of AI that can be used to create realistic images, videos, or audio recordings. Deepfake technology can manipulate an existing photo, video, or audio recording to switch one person for another. Deepfake technology can also create entirely new content in which someone appears to do or say something he or she did not.

To make a deepfake, AI algorithms train on many images and videos of a person (called the source) to learn the person's facial characteristics. The AI system learns to map that person's fa-

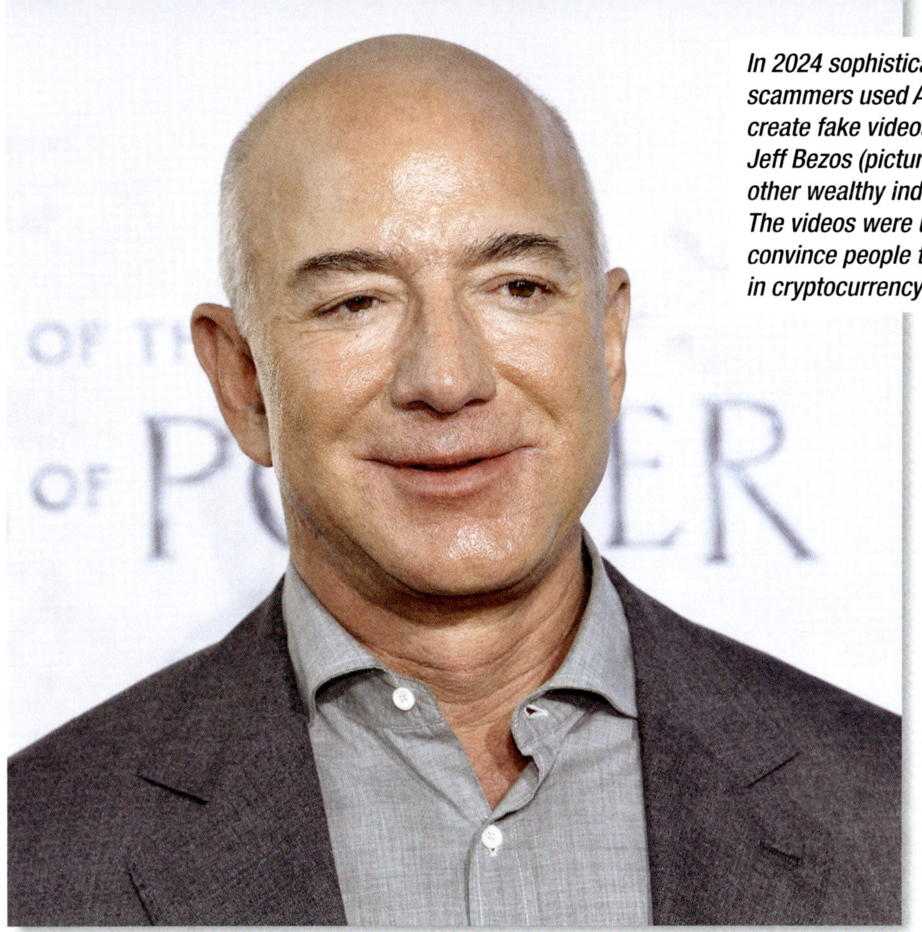
In 2024 sophisticated scammers used AI to create fake videos of Jeff Bezos (pictured) and other wealthy individuals. The videos were used to convince people to invest in cryptocurrency scams.

cial features onto another image or video (called the target). The AI system can then replace the face in the target video with the source's facial features. The system swaps faces and adjusts expressions, gestures, and mouth movements to match what it has learned about the source. Deepfakes put people into images and videos when they were not really there or make audio recordings of words they never said.

## How Deepfakes Are Used

In some cases, deepfakes are used for legitimate purposes. Deepfakes can be used in movies and television to de-age an actor or add a deceased actor to a scene. For example, the 2019 film *The Irishman* used computer-generated imagery and deepfake technology to age and de-age actors Al Pacino and Robert De Niro. Deepfake technology is also used at the Dalí Museum in Florida to create a deepfake video of the late artist Salvador Dalí, who welcomes visitors and tells them about his life and art. Movies and video games use deepfake technology to copy and manipulate actors' voices for scenes that are difficult to shoot or when the actors are no longer available to record their voices. Deepfakes are also used to create satire or parody content, such as the 2025 deepfake video of Dwayne "the Rock" Johnson as the character Dora the Explorer. Viewers enjoy the content but know that it is not real.

However, deepfakes are also being used in malicious ways. Deepfakes have the potential to do a lot of damage because it is tough to spot them. Scammers have used deepfake technology to create photos or videos of a person engaging in illegal or inappropriate situations, such as doing drugs or engaging in sex acts. Sometimes the scammers try to extort the victim and threaten to release the video if not paid. Other times, they use the deepfake content to humiliate victims and damage reputations.

*Deepfakes are used in movie production. The 2019 film* The Irishman *used computer-generated imagery and deepfake technology to age and de-age actor Robert De Niro (shown in the poster).*

## Deepfake Detection

In 2024 OpenAI—creator of ChatGPT and the AI image generator DALL-E—released a new AI tool to detect deepfakes. The company announced plans to share its deepfake detector with a number of disinformation researchers so they could test it and make suggestions for improvement. Open AI's deepfake detector was able to correctly flag nearly 99 percent of deepfake images created by DALL-E. However, the detector struggled to identify AI-generated deepfake images produced by other AI tools. OpenAI and other tech leaders, such as Google and Meta, have joined an effort to develop a label or credential for digital content, including images, videos, audio, and other files. The credential would show when and how content was produced and altered, including whether AI was used.

## Causing Harm

In 2023 twenty-two-year-old Patrick Carey from Long Island, New York, was sentenced to six months in jail after pleading guilty to a deepfake scheme. As part of this scheme, Carey copied school photos from the social media accounts of multiple women. He used deepfake technology to alter the photos and make them appear as if the women were engaging in sex acts. Then Carey posted the fake photos on a porn website, along with the women's names, addresses, and phone numbers. He encouraged others on the porn site to harass and threaten the women. "Patrick Carey targeted these women, altering images he took from their social media accounts and the accounts of their family members and manipulating them using 'deepfake' technology to create pornography that he disseminated across the Internet,"[35] Nassau County district attorney Anne Donnelly said in a statement.

In Maryland an athletic director was accused of using deepfake technology to harm the reputation of a school principal. Police alleged that Dazhon Darien, Pikesville High School's athletic director, used AI in January 2024 to create fake audio recordings of the school's principal, Eric Eiswert. In the fake audio, Eiswert appeared to make racist and anti-Semitic comments. The audio clip was sent in an email to a few of the school's teachers and spread on social media.

When the deepfake audio was made public, the backlash was swift. Although Eiswert insisted it was fake, he was placed on leave. Hate-filled messages and threats flooded social media, while angry callers phoned the school. In the investigation, forensic experts confirmed the audio had been faked, and police traced it back to Darien. Police believe Darien created the deepfake audio to attack Eiswert after the two men had discussed Darien's poor work performance and alleged misuse of school funds. The incident illustrated how easy it is for the average person to use AI technology with malicious intent. "Everybody is vulnerable to attack, and anyone can do the attacking,"[36] says Hany Farid, a professor at the University of California, Berkeley, who focuses on digital forensics and misinformation.

**"Everybody is vulnerable to attack, and anyone can do the attacking."[36]**

—Hany Farid, professor at the University of California, Berkeley

## Fraud and Financial Scams

Deepfake technology is a powerful tool in the hands of criminals and scammers. Deepfakes are increasingly being used as a tool for financial fraud. According to the FBI's Internet Crime Complaint Center, nearly 40 percent of financial fraud victims in 2023 were conned by schemes involving deepfake videos created with AI technology. The scams include investment scams, romance scams, and more. They also used AI to clone the voices of family, friends, and coworkers to trick people into sending money.

Financial scams cost Americans billions of dollars each year. In 2023 Americans lost approximately $12.5 billion from online scams, according to the FBI. Most of that money was never returned, according to Social Catfish, a company that helps people combat online crimes. Most online scammers live in other countries and request payment in cryptocurrencies, which are difficult to recover.

Online crime experts warn that deepfake scams will become more common as AI technology improves and becomes more chal-

## NO FAKES Act

A bipartisan group of US senators and representatives introduced the NO FAKES Act in Congress in 2024. The act aims to protect individuals from having their identity used in AI and deepfake technology without permission. The act would ban the unauthorized use of a person's voice, image, or likeness for AI or deepfake tools. The act is supported by many celebrities whose likenesses have been used in deepfake scams. Actor and comedian Steve Harvey, whose image, voice, and likeness have been stolen for this type of scam, supports the proposed legislation. "I prided myself on my brand being one of authenticity, and people know that, and so they take the fact that I'm known and trusted as an authentic person, pretty sincere," Harvey says.

Critics of the act, including the American Library Association and Electronic Frontier Foundation, warn that it could threaten First Amendment rights to free speech because it currently does not include any fair use provisions or creative expression protections. Harvey disagrees. "It's freedom of speech, it's not freedom of, 'make me speak the way you want me to speak.' That's not freedom, that's abuse," he says. As of April 2025, the act was under consideration in Congress.

Quoted in Hadas Gold, "Celebrity AI Deepfakes Are Flooding the Internet. Hollywood Is Pushing Congress to Fight Back," CNN, March 8 2025. www.cnn.com.

lenging to detect. "This national crisis is likely to get worse in the years ahead as scammers can now use artificial intelligence to create deepfake videos of business leaders, celebrities, politicians, and romantic suitors that are difficult to detect,"[37] says Social Catfish.

## Targeting Businesses

Businesses are also vulnerable to deepfake financial scams. In 2024 a finance worker at Arup, a British multinational design and engineering firm, fell victim to deepfake technology that impersonated the company's chief financial officer (CFO). The worker received a video meeting invitation, supposedly from the company's CFO. When the worker joined the video call, he recognized several colleagues. They looked and sounded like people he knew in real life. The worker did not realize that the video meeting was part of an elaborate scam. "(In the) multi-person video conference, it turns

*Businesses have been scammed by deepfakes. In one scam, during a video meeting, fake AI executives directed a subordinate to transfer millions of dollars to offshore accounts*

out that everyone [he saw] was fake,"[38] says senior superintendent Baron Chan Shun-ching with the Hong Kong police. Not realizing anything was wrong, the worker agreed when the CFO requested that he transfer $25.6 million to several bank accounts in Hong Kong. The worker later discovered the fraud after following up with Arup headquarters, and the company notified Hong Kong police.

As generative AI tools are used more frequently for fraud and scams, people should be on guard for potentially fake content in their personal and professional lives. Matthew Miller, a principal of cyber security services at KPMG, an accounting, tax, and consulting firm, warns individuals to view everything they see online with skepticism. "The public needs to maintain continuous vigilance when interacting through digital channels. Situational awareness and common sense go a long way to prevent an incident from occurring. If something does not 'feel right,' it has a high probability of being malicious activity,"[39] he says.

## Influencing Elections

Deepfakes have also been used to spread false information to influence elections and sway public opinion. Thousands of voters in

New Hampshire experienced this firsthand in January 2024 when they answered a phone call allegedly from President Joe Biden. On the call, Biden's voice asked Democrats not to vote in the New Hampshire primary, which was only a few days away. "We know the value of voting Democratic when our votes count. It's important you save your vote for the November election,"[40] the caller said. Recorded calls from politicians are common during election season. But this call was not from Biden. It was a deepfake created by generative AI technology.

It was later revealed that the deepfake was created by a Democratic political consultant who wanted to highlight the dangers of AI during an election season. The consultant wanted to show how easy it was to create and deploy a deepfake to influence voters and potentially impact the election's outcome. "The nightmare situation was the day before, the day of election, the day after election, some bombshell image, some bombshell video or audio would just set the world on fire,"[41] says Farid.

> "Deepfakes are especially concerning, not only because they're on the rise, but also because people are increasingly unsure of what's real and what's fake online."[43]
>
> —Abhishek Karnik, McAfee's head of threat research

While no bombshell deepfake content erupted during the 2024 elections, numerous deepfake memes circulated, often shared by politicians and their supporters. "I don't think the images were designed to be clearly deceptive, but they were designed to push a narrative, and propaganda works,"[42] says Farid. Sixty-three percent of Americans reported seeing deepfaked political memes in the two months leading up to the November 2024 presidential elections, according to a survey by the computer security company McAfee. Nearly half of those Americans admitted that the deepfake had influenced who they planned to vote for in the election. "Deepfakes are especially concerning, not only because they're on the rise, but also because people are increasingly unsure of what's real and what's fake online,"[43] says Abhishek Karnik, McAfee's head of threat research.

CHAPTER FIVE

# Synthetic Identity Fraud

In April 2024 Toronto police announced the arrest of twelve individuals alleged to have taken part in a wide-ranging synthetic identity fraud scheme. The two-year investigation began after a financial institution contacted police and reported several synthetic, or fake, accounts. Police say these accounts were created by blending stolen personal information with fabricated details. Those new synthetic identities were then used to open accounts, take out loans, and commit other financial crimes. The investigation uncovered a sophisticated fraud scheme that extended back to 2016 and involved hundreds of synthetic identities and multiple financial institutions. "The perpetrators of this scheme, which began in 2016, are alleged to have created more than 680 unique synthetic identities, many of which were used to apply for and open hundreds of bank and credit accounts at various banks and financial institutions across Ontario,"[44] said Toronto's Financial Crimes Detective David Coffey at a press conference.

Using the fraudulently obtained accounts, the thieves were able to make in-store and online purchases, withdraw cash, and initiate electronic fund transfers. Toronto police estimated that the thieves stole at least $4 million. They warned that the amount could rise as other victims were identified.

Sophisticated technology played a role in the synthetic identity scheme. As part of their investigation, Toronto police seized dozens of fake government IDs and the electronic templates used to create them. They also seized hundreds of debit and credit cards linked to the synthetic identities. Experts believe that AI likely played a role in the scheme, helping the perpetrators scan the internet for personal data, generate dozens of fake identities, and send hundreds of credit applications to financial institutions.

## What Is Synthetic Identity Fraud?

Identity theft is not a new problem. Criminals steal an individual's personal information and use it for financial gain. Personal information includes names, addresses, and Social Security numbers. It can also include account numbers for credit cards, medical insurance, and bank accounts. Using these details, thieves are able to obtain credit cards and loans. They have also found ways to obtain medical insurance payments. "Synthetic identity fraud involves creating an entirely new person who doesn't exist legally by combining stolen, manipulated and fictitious personally identifiable information along with bogus facial imprints,"[45] says Shahid Hanif, founder of Shufti Pro, a biometric identity verification company. Because of its combination of real and fake information, synthetic identity fraud is sometimes called "Frankenstein fraud."

> **"Synthetic identity fraud involves creating an entirely new person who doesn't exist legally by combining stolen, manipulated and fictitious personally identifiable information along with bogus facial imprints."[45]**
>
> —Shahid Hanif, founder of Shufti Pro, a biometric identity verification company

Synthetic identity fraud is often associated with financial fraud targeting financial institutions, but it can be used in other ways. Criminals use synthetic identities to enroll in government benefit and service programs. For example, two Florida men were charged in 2020 with using synthetic identities to steal millions of dollars from federal COVID-19 relief programs. Synthetic identities can also be used to create a fake business or apply for various licenses, which make the fictional identity appear more real. The more realistic the synthetic identity appears, the easier it becomes for criminals to use it. Jeffrey Huth, a senior vice president at the credit reporting agency

***Using synthetic identity fraud, scammers can create fake identities and then use them to obtain credit and debit cards.***

TransUnion, warns that everyone should be on the lookout for AI-generated synthetic identities. "I think it will become much easier and faster to create completely realistic-looking, fabricated identities, whether it's building a financial profile, [or] a digital footprint,"[46] he says.

## Fastest-Growing Financial Fraud

Synthetic identity fraud has become the fastest-growing form of financial fraud in the United States, according to a 2024 report by TransUnion. The report noted that incidents of synthetic fraud increased by about 184 percent from 2019 to 2023. Fraud experts point to AI as one of the primary reasons for the surge in synthetic financial fraud. "It's easier and easier for people to create synthetic identities. Using either stolen information or made-up information using generative AI,"[47]says Andrew Davies, of the global technology firm ComplyAdvantage.

> "I think it will become much easier and faster to create completely realistic-looking, fabricated identities, whether it's building a financial profile, [or] a digital footprint."[46]
>
> —Jeffrey Huth, a senior vice president at TransUnion

Synthetic identity fraud has become increasingly popular among criminals because of the ability to create an unlimited number of identities that are very challenging to detect. Efforts to stop synthetic identity fraud are made even more difficult by the speed with which fake identities can be created and the realism that can be achieved. Banks "lose billions every year from the activity of fake identities," says Ari Jacoby, chief executive officer of cybersecurity firm Deduce. Jacoby continues:

> This often happens in the form of account fraud—opening up lines of credit under a fake identity that the fraudster leverages before disappearing. And it's only getting worse thanks to AI, which makes it quicker and easier than ever to create fake identities. AI also opens up the floodgates to more convincing audio and video deepfakes—which . . . have already duped a third of businesses worldwide.[48]

## Using AI to Create a New Identity

Stealing personal information—whether through a data breach, malware, or other means—is often the first step in creating fake identities. Social Security numbers, which are assigned to nearly all US citizens and some noncitizens, are a key piece of information used in creating synthetic identities. AI is the tool of choice for collecting other information that helps criminals create those identities. AI tools scan and collect massive amounts of personal data obtained through breaches as well as from public databases and social media. Generative AI combs through all of this data, mixing and matching details and inventing new ones to create multiple synthetic identities. Generative AI can quickly develop a convincing biography and work history to make each invented identity appear legitimate.

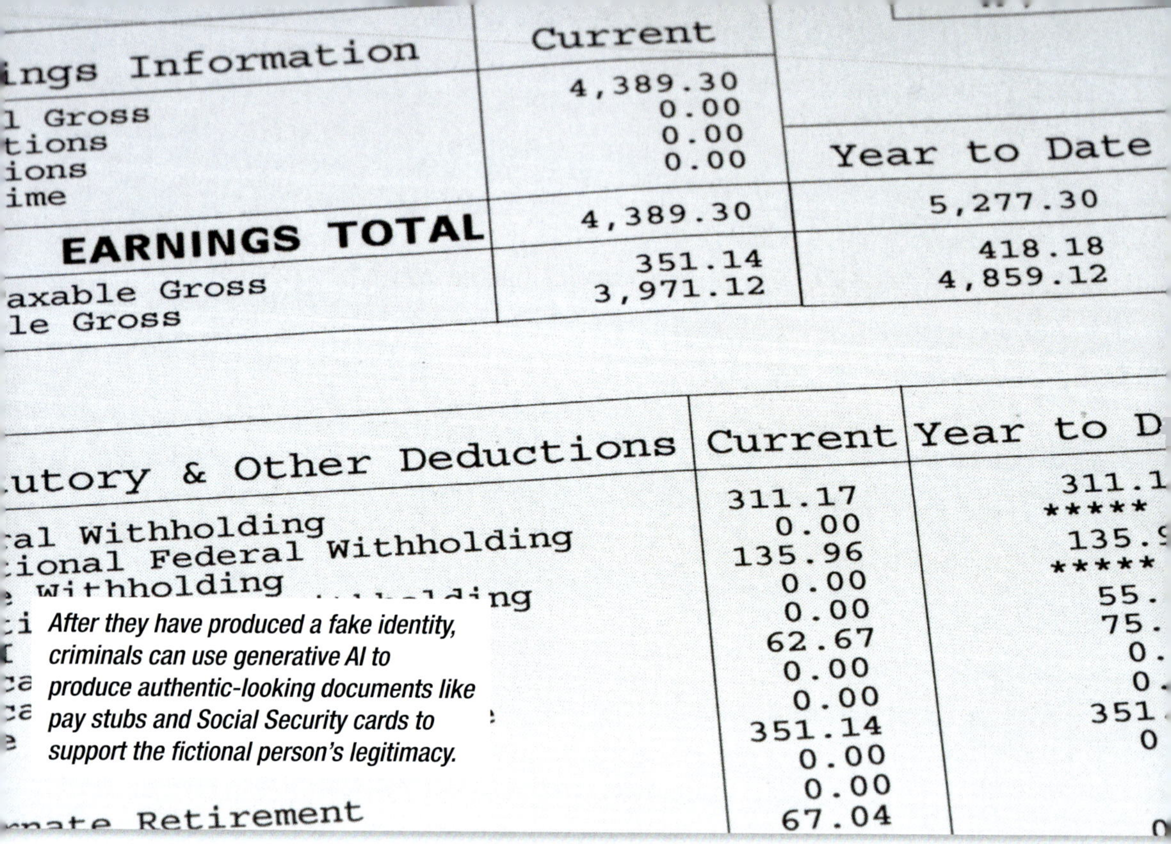

*After they have produced a fake identity, criminals can use generative AI to produce authentic-looking documents like pay stubs and Social Security cards to support the fictional person's legitimacy.*

Once criminals create a fake identity, they can use generative AI to produce authentic-looking documents such as birth certificates, Social Security cards, pay stubs, bank statements, and utility bills that support the legitimacy of the fictional person. Generative AI can also use photos found online to create a realistic driver's license for the fictional person. For example, a criminal can simply prompt a generative AI tool to create a driver's license from a given state, and it will be able to create the license using pictures of people found online, Jacoby says. Generative AI can also create deepfake images and videos that can be used to set up facial recognition access on credit card and bank accounts opened for the fake identity.

## Using Synthetic Identities for Fraud

AI technology can also make it easy for criminals to flood banks, credit card companies, and other businesses with bogus applications for accounts, loans, and other services. AI tools can even use the same synthetic identities to apply for credit and open accounts at multiple financial institutions simultaneously. Aside from these abilities, AI technology has yet another benefit: it learns from failures and adapts. For example, if a credit card company turns down an application from one synthetic identity, AI can change a few details for the next application. With each application, the AI system learns how to improve. According to synthetic fraud experts at the Federal Reserve:

> These Gen AI features increase the number of accounts opened with synthetic identities that then can be used to steal money by making credit card purchases or overdrawing an account with no intent to repay. Fraudsters may choose to let accounts mature to obtain higher credit limits or apply for loans to increase the amount available to steal. The accounts can be used to receive and transfer money to support other criminal activities, such as fraud, scams, and money laundering.[49]

Jacoby warns that the accessibility of AI tools allows criminals who are not technically sophisticated to use them in their fraud schemes. "There's an enormous group of bad guys, bad folks out there, that are now weaponizing this type of artificial intelligence to accelerate the pace at which they can commit crimes. That's the low end of the spectrum. Imagine what's happening on the high end of the spectrum with organized crime and enormous financial resources,"[50] he says.

Many cybersecurity experts believe that AI-fueled synthetic identity fraud will become an even bigger problem in the future. The Deloitte Center for Financial Services estimates that synthetic identity fraud could cause $23 billion in losses by 2030. Shahid Hanif says:

> Before the surge in digitization, identity theft was restricted to one person presenting a stolen identity at a time. The

fraud process was slow, and the intelligence of in-house employees was key to detecting and mitigating such fraud risks. With the increase in the creation of synthetic IDs . . . cybercriminals can defraud businesses and individuals on a large scale. Using synthetic identities, criminals can produce countless identities that they can impersonate to gain unauthorized access. Using AI, potentially hundreds of these attacks can be launched from any part of the world to reap illicit benefits.[51]

## Detection Challenges

As AI technology advances, synthetic identity fraud will become even more difficult to detect. Improvements in AI have the potential to make it even more difficult to tell the difference between what is real and what is fake, enabling criminals to be more effective with their scams.

"AI-generated . . . synthetic identities are the bad actors' newest tool in defrauding financial institutions, but the widespread impacts don't stop there. Given that synthetic identity fraud was already the most common form of identity fraud in the U.S., the prospect of fraudsters becoming more agile and effective is a sobering one,"[52] says Jacoby.

**"AI-generated . . . synthetic identities are the bad actors' newest tool in defrauding financial institutions."[52]**

—Ari Jacoby, chief executive officer of cybersecurity firm Deduce

Typically, financial institutions rely on machine learning models to flag suspicious accounts and transactions. These AI models quickly analyze large amounts of data to identify transactions and patterns that could indicate fraud and flag them for further investigation. These machine learning algorithms become more accurate and effective over time. However, many of the existing tools used to detect fraud are insufficient for handling sophisticated AI-generated synthetic identities. In fact, 85 percent of synthetic identities were not flagged by fraud risk models in a 2021 study by LexisNexis Risk

## Property Fraud

Criminals are increasingly using AI tools in property fraud. Scammers use AI tools and deepfake technology to impersonate real estate agents, buyers, and sellers and to generate fake property documents. In one common scheme, criminals impersonate property owners and create fake documents to sell a property they do not own. One example occurred in Florida in September 2024. A woman contacted Florida Title and Trust, a company that assists in real estate transactions, to say she wanted to sell a vacant lot.

But something about the call did not seem right, leading the title company to request a video call to verify the woman's identity. During the call with the seller, title company owner Lauren Albrecht quickly realized she was looking at a deepfake video set to replay on a loop. When Albrecht asked the woman to raise her hand, the woman in the video never reacted. After the call, Albrecht used a reverse image search and matched the seller's identification with a photo of a California woman who had been reported missing years earlier. Albrecht was relieved that she uncovered the scam before any real damage was done. However, she wonders how many other real estate professionals have been fooled by similar AI-powered property scams.

Solutions, a global data and analytics company that provides fraud prevention services.

## Advanced Biometrics Solutions

Some companies are exploring advanced biometrics solutions to combat synthetic identity fraud. Biometrics are measurable physical characteristics or behaviors that can be used to verify a person's identity. While traditional biometric tests such as facial images, fingerprints, and iris scans have been deceived by AI-generated deepfakes and synthetic identities, advanced biometric tools have the potential to better detect fraud.

One area of promise involves tools that can verify that an institution is interacting with a live individual instead of an AI-generated deepfake in, for example, a video call. These so-called liveness detection tools use various tests to verify whether a person is real or fake. "These tests may use a range of techniques to verify that a user is responding in real time, for example, by asking them to tilt their head to the side, smile, or blink,"[53] write representatives of Deloitte Center for Financial Services. Advanced physical biometric tools can also verify a person's identity by examining physical features such as skin texture, facial imperfections, and perspiration. Other tools can even detect whether a person's blood is flowing.

Developing behavioral biometric tools may also be helpful in combating synthetic identity fraud. Behavioral biometric tools track how a person types on a computer, moves a mouse, and applies pressure to a touch screen. Cyber experts predict that behavioral biometrics will be very effective at flagging synthetic identities. This is because fake identities often display different patterns than live humans do when filling out multiple forms across websites.

*Liveness detection tools can verify that an institution is interacting with a live individual instead of an AI-generated deepfake, for example, in a video call.*

## Social Engineering and AI

Social engineering attacks attempt to manipulate people into sharing personal information and engaging in other unsafe online behavior. Cybercriminals often use social engineering to steal personal or financial data, including Social Security numbers, log-in passwords, bank account numbers, and more that they can use for identity theft. Generative AI is transforming social engineering attacks. Cybercriminals can use AI to create convincing images, text, audio, and video to scam and steal personal or financial data. For example, AI-generated emails, websites, and text messages deliver phishing attacks designed to trick unsuspecting victims into revealing sensitive personal data or organizational information. Generative AI can produce authentic-sounding phishing emails and fake social media profiles used to trick victims into sending money. Using AI is faster than writing emails and creating profiles manually. As a result, cybercriminals can automate and distribute their attacks faster, allowing them to trick more people with their scams.

AI is powering synthetic identity fraud, helping criminals create highly realistic fake identities with generative AI, deepfake technology, and stolen personal data. Synthetic identities are often able to pass undetected through traditional security measures and have the potential to cause financial institutions significant losses. To combat AI-powered synthetic identities, organizations are working to develop advanced biometric tools that will be better able to tell the difference between what is real and what is fake.

## Introduction: A Realistic Deception

1. Quoted in Carlos Granda, "Fraudsters Use Voice-Cloning AI to Scam Man Out of $25,000," ABC7 Los Angeles, October 18, 2024. https://abc7.com.
2. Quoted in Granda, "Fraudsters Use Voice-Cloning AI to Scam Man Out of $25,000."
3. Quoted in Granda, "Fraudsters Use Voice-Cloning AI to Scam Man Out of $25,000."
4. Quoted in Granda, "Fraudsters Use Voice-Cloning AI to Scam Man Out of $25,000."
5. Quoted in Katie J. Skipper, "Voice Cloning AI Scams Are on the Rise," BECU, September 27, 2024. www.becu.org.
6. Quoted in Federal Bureau of Investigation, "FBI Warns of Increasing Threat of Cyber Criminals Utilizing Artificial Intelligence," May 8, 2024. www.fbi.gov.

## Chapter One: What Is Artificial Intelligence?

7. Quoted in Florence Gonsalves et al., "AI—the Good, the Bad, and the Scary," *Virginia Tech Engineer*, Fall 2023. https://eng.vt.edu.
8. Quoted in Kate O'Sullivan, "AI Expert Spotlight: Georgia Perakis," MIT Sloan, October 31, 2024. https://mitsloan.mit.edu.
9. Quoted in Gonsalves et al., "AI—the Good, the Bad, and the Scary."
10. Quoted in Gonsalves et al., "AI—the Good, the Bad, and the Scary."
11. Rohit Kundu, "AI Risks: Exploring the Critical Challenges of Artificial Intelligence," Lakera, March 19, 2024. www.lakera.ai.

## Chapter Two: Plagiarism and Cheating

12. Quoted in Ben Moore, "'I Used AI to Cheat at Uni and Regret It,'" BBC, October 24, 2024. www.bbc.com.
13. Quoted in Moore, "'I Used AI to Cheat at Uni and Regret It.'"
14. Maddy Osman, "Is It Plagiarism to Use AI-Generated Content? The Ethics of Content Creation in an AI World," *The Blogsmith Blog*, July 17, 2024. www.theblogsmith.com.
15. Quoted in Beatrice Nolan, "AI Plagiarism Is Spreading in US Colleges. It's Left Professors Feeling Confused and Exhausted," Yahoo! Tech, September 30, 2024. www.yahoo.com.
16. Quoted in Pia Ceres and Amanda Hoover, "Kids Are Going Back to School. So Is ChatGPT," *Wired*, August 23, 2023. www.wired.com.
17. Quoted in Rachel Hale, "She Lost Her Scholarship over an AI Allegation—and It Impacted Her Mental Health," *USA Today*, January 22, 2025. www.usatoday.com.

18. Quoted in Colin Wood, "AI Detectors Are Easily Fooled, Researchers Find," EdScoop, September 10, 2024. https://edscoop.com.
19. Quoted in Wood, "AI Detectors Are Easily Fooled, Researchers Find."
20. Quoted in Junior Achievement USA, "Back to School Survey: 44% of Teens Likely to Use AI to Do Their Schoolwork for Them This School Year," 2023. https://jausa.ja.org.
21. Quoted in BBC News, "'Most of Our Friends Use AI in Schoolwork,'" October 31, 2023. www.bbc.com.

## Chapter Three: Misinformation and Manipulation

22. Quoted in Catherine Thorbecke, "Plagued with Errors: A News Outlet's Decision to Write Stories with AI Backfires," CNN, January 25, 2023. www.cnn.com.
23. Quoted in Paul Farhi, "A News Site Used AI to Write Articles. It Was a Journalistic Disaster," *Washington Post*, January 17, 2023. www.washingtonpost.com.
24. Quoted in Dan Milmo, "Google AI Chatbot Bard Sends Shares Plummeting After It Gives Wrong Answer," *The Guardian* (Manchester, UK), February 9, 2023. www.theguardian.com.
25. Quoted in Milmo, "Google AI Chatbot Bard Sends Shares Plummeting After It Gives Wrong Answer."
26. Quoted in Julia Busiek, "Three Fixes for AI's Bias Problem," University of California, March 21, 2024. www.universityofcalifornia.edu.
27. Quoted in Mike Allen, "AI and the Spread of Fake News Sites: Experts Explain How to Counteract Them," Virginia Tech, February 22, 2024. https://news.vt.edu.
28. Quoted in Pranshu Verma, "The Rise of AI Fake News Is Creating a 'Misinformation Superspreader,'" *Washington Post*, December 17, 2023. www.washingtonpost.com.
29. US Cyber Command, "Russian Disinformation Campaign 'DoppelGänger' Unmasked: A Web of Deception," 2024. www.cybercom.mil.
30. Quoted in Shannon Bond, "AI-Generated Spam Is Starting to Fill Social Media. Here's Why," NPR, May 14, 2024. www.npr.org.
31. Quoted in Bond, "AI-Generated Spam Is Starting to Fill Social Media."
32. Quoted in Verma, "The Rise of AI Fake News Is Creating a 'Misinformation Superspreader.'"

## Chapter Four: Deepfakes

33. Quoted in New York State Attorney General, "Attorney General James Warns New Yorkers of Investment Scams Using AI-Manipulated Videos," August 29, 2024. https://ag.ny.gov.
34. Quoted in Brian New and Lexi Salazar, "Deepfakes of Elon Musk Are Contributing to Billions of Dollars in Fraud Losses in the U.S.," CBS News, November 20, 2024. www.cbsnews.com.

35. Quoted in Pei-Sze Cheng and Jennifer Millman, "Long Island Man Jailed in Deepfake Sex Scheme Targeting 11 Women from His High School," NBC 4 New York, April 18, 2023. www.nbcnewyork.com.
36. Quoted in Ben Finley, "Deepfake of Principal's Voice Is the Latest Case of AI Being Used for Harm," AP News, April 29, 2024. https://apnew .com.
37. Quoted in Russ Wiles, "AI 'Deepfake' Videos Make Investment Scams Harder to Spot as Americans Lose Billions," *USA Today*, October 30, 2024. www.usatoday.com.
38. Quoted in Heather Chen and Kathleen Magramo, "Finance Worker Pays Out $25 Million After Video Call with Deepfake 'Chief Financial Officer,'" CNN, February 4, 2024. www.cnn.com.
39. Quoted in Jason Nelson, "AI Deepfakes Are a Threat to Businesses Too—Here's Why," Decrypt, August 25, 2023. https://decrypt.co.
40. Quoted in Shannon Bond, "How AI Deepfakes Polluted Elections in 2024," NPR, December 21, 2024. www.npr.org.
41. Quoted in Bond, "How AI Deepfakes Polluted Elections in 2024."
42. Quoted in Bond, "How AI Deepfakes Polluted Elections in 2024."
43. Quoted in Daniella Genovese, "Nearly 50% of Voters Say Deepfakes Had Some Influence on Election Decision: Survey," FOX 9 Minneapolis-St. Paul, October 31, 2024. www.fox9.com.

## Chapter Five: Synthetic Identity Fraud

44. Quoted in Ron Fanfair, "Arrests in Synthetic Identity Fraud," Toronto Police Service, April 29, 2024. www.tps.ca.
45. Shahid Hanif, "Council Post: Synthetic Identities: The Darker Side of Generative AI," *Forbes*, August 12, 2024. www.forbes.com.
46. Quoted in Sophia Fox-Sowell, "Synthetic Identity Fraud Is on the Rise, TransUnion Report Shows," StateScoop, June 26, 2024. https://state scoop.com.
47. Quoted in Ellen Sheng, "Generative AI Financial Scammers Are Getting Very Good at Duping Work Email," CNBC.com, February 14, 2024. www.cnbc.com.
48. Ari Jacoby, "Fake People Ruining the Real World? Finance Might Have an AI Problem," *Forbes*, February 1, 2024. www.forbes.com.
49. Federal Reserve, "Generative Artificial Intelligence Increases Synthetic Identity Fraud Threats." https://fedpaymentsimprovement.org.
50. Quoted in Tatiana Walk-Morris, "Why Criminals Like AI for Synthetic Identity Fraud," Dark Reading, March 5, 2024. www.darkreading.com.
51. Hanif, "Council Post."
52. Jacoby, "Fake People Ruining the Real World?"
53. Satish Lalchand et al., "Using Biometrics to Fight Back Against Rising Synthetic Identity Fraud," Deloitte Insights, July 27, 2023. www2 .deloitte.com.

**Association for the Advancement of Artificial Intelligence (AAAI)**
https://aaai.org
The AAAI is a nonprofit scientific society devoted to advancing the scientific understanding of AI and promoting its responsible use. Its website features magazine articles and press about AI, as well as information about industry conferences and symposia.

**Brookings Institution**
www.brookings.edu
The Brookings Institution is a nonprofit public policy organization based in Washington, DC. It strives to conduct nonpartisan, in-depth research on problems facing society at the local, national, and global level, including artificial intelligence. Its website features many articles about AI.

**Cato Institute**
www.cato.org
The Cato Institute is a public policy research organization that strives to promote libertarian ideas of individual liberty, limited government, and free markets. A search of its website returns several articles on artificial intelligence topics.

**Center for AI Safety (CAIS)**
www.safe.ai
The CAIS seeks to reduce AI risks through research, funding, and advocacy. The nonprofit organization aims to identify and address AI safety issues before they become significant concerns. Its website features information about AI risks, research projects, and more.

**OpenAI**
https://openai.com
OpenAI is an AI research and deployment company that aims to ensure that artificial intelligence benefits all of humanity. It is the developer of ChatGPT, a generative AI technology. Its website has information about the company and its research. Users can test ChatGPT on the website.

**Partnership on AI**
https://partnershiponai.org
The Partnership on AI is a nonprofit organization that promotes the responsible advancement and use of AI technology. Its website features articles, a blog, and other information about AI and its use.

# FOR FURTHER RESEARCH


## Books

John Allen, *Artificial Intelligence: Promise and Peril*. San Diego, CA: ReferencePoint, 2024.

Lisa Idzikowski, *Artificial Intelligence and the Future of Humanity*. New York: Greenhaven, 2023.

Stuart A. Kallen, *Changing Lives Through Artificial Intelligence*. San Diego, CA: ReferencePoint, 2021.

Cade Metz, *Genius Makers: The Mavericks Who Brought AI to Google, Facebook, and the World*. New York: Dutton, 2021.

Carla Mooney, *Artificial Intelligence*. San Diego, CA: ReferencePoint, 2025.

Scientific American Educational, *Artificial Intelligence*. New York: Scientific American Educational, 2023.

## Internet Sources

Associated Press, "Paper Exams, Chatbot Bans: Colleges Are Looking to 'ChatGPT-Proof' Assignments," ABC13, August 19, 2023. https://abc13.com.

Devin Coldewey, "Age of AI: Everything You Need to Know About Artificial Intelligence," TechCrunch, August 4, 2023. https://techcrunch.com.

David H. Freedman, "How AI Will Make Our Lives Better (and Worse)," *Newsweek*, May 31, 2023. www.newsweek.com.

Bernard Marr, "The Difference Between Generative AI and Traditional AI: An Easy Explanation for Anyone," *Forbes*, July 24, 2023. www.forbes.com.

Cade Metz, "What Exactly Are the Dangers Posed by A.I.?," *New York Times*, May 1, 2023. www.nytimes.com.

Adam Satariano and Paul Mozur, "The People Onscreen Are Fake. The Disinformation Is Real," *New York Times*, February 7, 2023. www.nytimes.com.

Stuart A. Thompson, "Making Deepfakes Gets Cheaper and Easier Thanks to A.I.," *New York Times*, March 12, 2023. www.nytimes.com.

*Note: Boldface page numbers indicate illustrations.*

Cover: Andrew Derr/Shutterstock

6: Rawpixel.com/Shutterstock
10: Elkins Eye Visuals/Shutterstock
12: Kaspars Grinvalds/Shutterstock
15: panuwat phimpha/Shutterstock
20: Ground Picture/Shutterstock
22: Sharaf Maksumov/Shutterstock
24: Jessica Jeong/Shutterstock
29: OpturaDesign/Shutterstock
31: mundissima/Shutterstock
34: Oleksii Synelnykov/Shutterstock
39: DFree/Shutterstock
40: kcube - kaan baytur/Shutterstock
44: fizkes/Shutterstock
48: Media/Shutterstock
50: Josh randall/Shutterstock
54: Fizkes/Shutterstock

## ABOUT THE AUTHOR

Carla Mooney is the author of many books for young adults and children. She lives in Pittsburgh, Pennsylvania, with her husband and three children.